Insects on Grain Legumes in Northern Australia

Insects on Grain Legumes in Northern Australia

A Survey of Potential Pests and Their Enemies

Merle Shepard,
R.J. Lawn
and
Margaret A. Schneider

University of Queensland Press
St Lucia • London • New York

Typeset by University of Queensland Press
Printed and bound by Warren Printing Co Ltd, Hong Kong

Distributed in the United Kingdom, Europe, the Middle East, Africa, and the Caribbean by Prentice-Hall International, International Book Distributors Ltd, 66 Wood Lane End, Hemel Hempstead, Herts., England.

Published with the assistance of the CSIRO
Division of Tropical Crops and Pastures.

National Library of Australia
Cataloguing-in-Publication data

Shepard, Merle, 1942–.
Insects on grain legumes in northern Australia.

Bibliography.
Includes index.
ISBN 0 7022 1802 2.

1. Legumes — Diseases and pests — Australia, Northern. 2. Insects, Injurious and beneficial — Australia, Northern. I. Lawn, R.J. (Robert John), 1945–. II. Schneider, Margaret A. (Margaret Anne), 1948–. III. Title.

633.3′0497′0994

Library of Congress Cataloging in Publication Data

Shepard, Merle.
Insects on grain legumes in northern Australia.

Bibliography: p.
Includes indexes.
1. Legumes — Diseases and pests — Australia.
2. Insects, Injurious and beneficial — Australia.
3. Insects — Australia — Parasites. 4. Insect control — Biological control — Australia. I. Lawn, R.J., 1945- . III. Schneider, Margaret A., 1948-
III. Title. IV. Title: Grain legumes in northern Australia.
SB608.L4S53 1983 633.3′0497′099429 82-13463

Library of Congress Cataloging in Publication Data

ISBN 0-7022-1802-2 (pbk.)

CONTENTS

ACKNOWLEDGEMENTS

The extent of the survey in Western Australia and the Northern Territory would have been impossibly restricted without the generous assistance and local knowledge volunteered by many people. In particular, the authors wish to thank Geoff Strickland, Alan Garside, and Stewart Learmoth of the Western Australian Department of Agriculture, Kununurra; Russel Muchow, Kimberley Research Station, Kununurra; Allan Allwood and Leon Radunz, Northern Territory Department of Primary Production, Darwin; and Stan Putland, Northern Territory Department of Primary Production, Katherine.

In Queensland, the assistance of Bruce Pyke, Grant Adams, and Peter Blood of the University of Queensland is acknowledged.

The authors also thank the following people for their assistance with identification of material: Ian Galloway, Bryan Cantrell, Ken Houston, and John Donaldson of Entomology Branch, Department of Primary Industries, Queensland; Ted Dahms, Valerie Davies, and Robert Raven of Queensland Museum; Ian Common, Elwood Zimmerman, and Tom Weir of the Australian National Insect Collection, CSIRO, Canberra; Tom Woodward and Greg Daniels of the University of Queensland; and Howard Evans of Colorado State University, USA.

Thanks also to Anne Ting for typing the manuscript during preparation of this booklet.

INTRODUCTION

The grain legumes are among the most susceptible of all crops to insect attack and around the world provide a nutritious and palatable substrate for a large and extremely diverse range of pests (van Emden 1981). Although production of grain legumes in northern Australia is small, the areas sown to these crops are expanding (Williamson 1971, Lawn and Russell 1978, Lawn and Byth 1979, Wood and Russell 1979). Environmental conditions in the tropics and sub-tropics especially favour rapid development of dense and diverse insect communities. Problems with insect pests can therefore be expected to increase with this expansion, particularly where previously unnoticed native species find the new crops to their liking.

One of the first steps toward development of economical and stable control programmes for insect pests in the grain legumes is the identification of the pest and beneficial species associated with these crops (e.g., see the approach by Carner, Shepard, and Turnipseed 1974*a* and 1974*b* and Shepard, Carner, and Turnipseed 1975 for soybeans in the south-eastern United States).

Several reports have documented the major current insect pests of grain legumes in Australia, particularly in subtropical and temperate regions (e.g., see the review by Turner 1978). Reports that consider both pests and beneficial species, however, are rare. Richards (1968) provided a comprehensive documentation of insects found in a number of crops, including several grain legume species, in the Ord River Irrigation Area. A sweep net survey of soybean insects in New South Wales revealed 145 species from 72 families (Miller, Rose, and McDonald 1979). However, that survey did not differentiate between pest and beneficial species, and many of the important pest species that occur in grain legumes in Queensland, the Northern Territory, and Western Australia were not reported.

Because of the importance of the subject, and the fragmentary nature of current information, it was decided to obtain and present survey information about the insect pests and their natural enemies associated with grain legumes in tropical and sub-tropical Australia. Colour plates of most of the potentially important economic species have been included to provide a ready field guide for use by farmers, extension workers, agronomists, entomologists, and others.

It is to be hoped the information in this booklet will contribute to a greater awareness of the importance of the natural enemies of insect pests in the development of economically and ecologically sound pest-management practices.

The current state of grain legume production in northern Australia provides an unusual opportunity for incorporating integrated pest-management principles during the development of new production systems. Integrated pest management combines the use of a range of control measures, including the conservation of natural control agents such as arthropod predators and parasites and insect diseases.

Research on integrated pest management in soybeans has shown conclusively that natural enemies can play an important role in preventing insect pests from causing economic damage (Turnipseed 1972, Newsom 1978). This has been demonstrated by using chemicals either to remove these natural enemies or to reduce their effectiveness. One consequence of insecticide use may be a resurgence of pest populations to densities far greater than those in untreated areas (Shepard, Carner, and Turnipseed 1977). In addition, previously minor pests may become serious when their natural enemies are killed off by chemicals.

Integrated pest management systems may not overcome the need for some pesticide use. Where pesticides are necessary, however, their use should be restricted to selected applications of chemicals that are less harmful to important beneficial species but effective in keeping pest populations under control (Turnipseed 1972).

SURVEY METHODS

Surveys were conducted during the period from October 1978 to April 1981 within selected grain legume fields in the Brisbane Valley, Lockyer Valley, Darling Downs, and Burnett regions of Queensland, in the Ord River Irrigation Area of Western Australia, and near Darwin and Katherine in the Northern Territory (see map). These areas included the major areas where grain legumes are currently grown with the exception of the Central Highlands and Atherton Tableland areas in Queensland. Information was gathered by walking through the fields and examining plants, with particular attention given to those with damaged leaves, stems, and pods. In some instances sweep nets and ground sheets were also used. The efficacy of various sampling techniques has been discussed by Shepard, Carner, and Turnipseed (1974) in relation to insect populations in soybeans in the south-eastern United States.

To determine the incidence of parasitism and disease of some caterpillar species, over three thousand collections were made and larvae raised on artificial diet. This allowed incidence of mortality to be determined and identification of beneficial agents to be made. When possible, pest and beneficial insects and spiders were collected and many of the pest species that could not be reared on artificial diet were reared on their host plants. This procedure allowed photographs of various growth stages to be taken and facilitated observation of natural parasites and diseases.

Information on the predacious habit of many species was obtained in studies using *Heliothis* spp. labelled with ^{32}P isotope in soybean fields in south-eastern Queensland. Many naturally occurring predators of both the eggs and larvae of *Heliothis* spp. were recorded. Similar studies in South Carolina found that a total of 349 arthropod predators from twenty-five families fed on the eggs and larvae of several important caterpillar pests of soybean (McCarty, Shepard, and Turnipseed 1980).

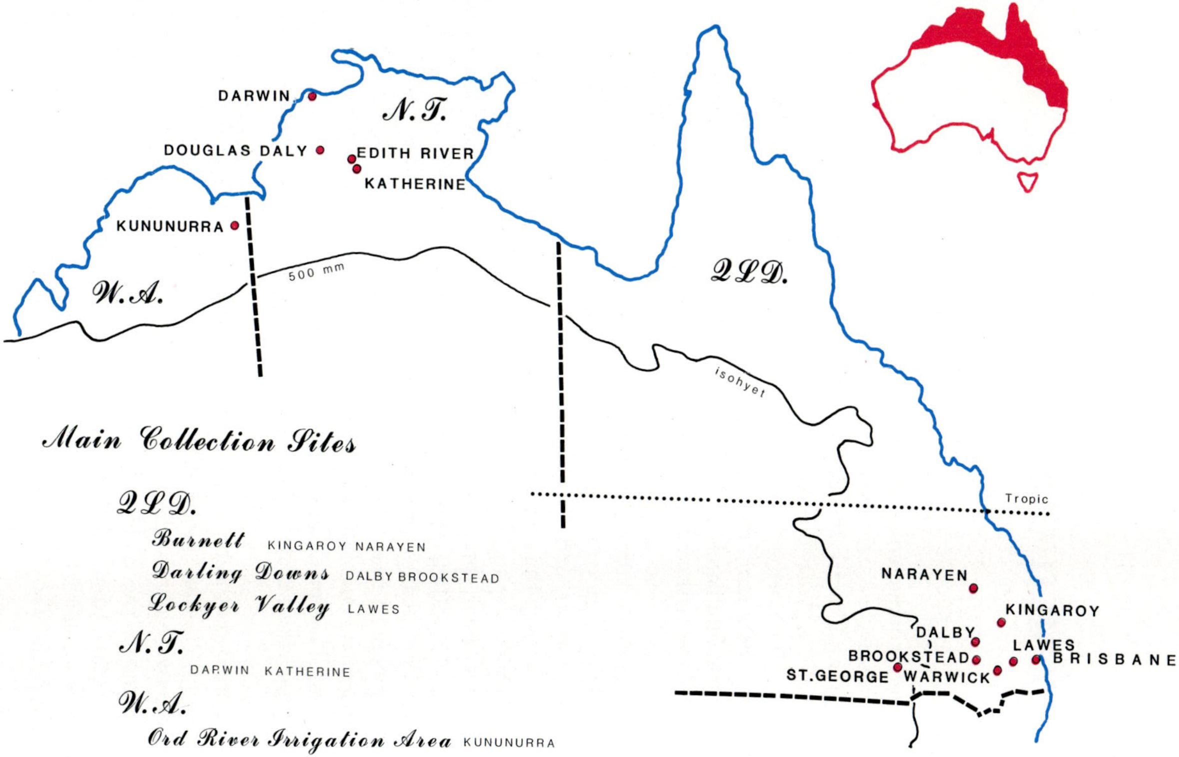
DARWIN
N.T.
DOUGLAS DALY
EDITH RIVER
KATHERINE
KUNUNURRA
W.A.
500 mm
QLD.
isohyet
Tropic
NARAYEN
KINGAROY
DALBY
LAWES
BROOKSTEAD
BRISBANE
ST.GEORGE
WARWICK
Main Collection Sites
QLD.
Burnett KINGAROY NARAYEN
Darling Downs DALBY BROOKSTEAD
Lockyer Valley LAWES
N.T.
DARWIN KATHERINE
W.A.
Ord River Irrigation Area KUNUNURRA

Insects such as purely nectar feeders and feeders on plant debris were ignored. The main grain legume crops surveyed were soybean (*Glycine max*), mungbean (black gram, *Vigna mungo;* green gram, *V. radiata*), cowpea (*V. unguiculata*), pigeonpea (*Cajanus cajan*), peanut (*Arachis hypogaea*), and lablab bean (*Lablab purpureus*). All are currently summer or wet season crops. Not all crops were present in all regions surveyed, and some summer-grown grain legumes (e.g. navy beans [*Phaseolus vulgaris*]) were not surveyed.

OUTLINE OF SURVEY RESULTS

Over 260 arthropod species were recorded in the survey as having either potential pest or beneficial status. These are tabulated in index 1 (pests) and index 2 (beneficials). Approximately half of all species were potential pests. There are colour plates for 90 species, and brief notes are provided on some aspects of the biology of those species currently recognized as economically important. However, the biology and potential economic importance of many of the species are not known. About one-third of all species recorded could not be fully identified, and many of these species remain undescribed.

The four insect orders that contained most potential pest species recorded were the bugs (Hemiptera; 36 per cent cent), moths and butterflies (Lepidoptera; 28 per cent), beetles (Coleoptera; 20 per cent), and grasshoppers and crickets (Orthoptera; 13 per cent). In this booklet the various species are discussed on the basis of damage to the crop.

In general, the pest complexes occurring on the grain legumes in the different geographical locations were similar. The presence of individual species, however, and their relative abundance varied with location and species of grain legume. For example, a pod-sucking bug complex was ubiquitous, and some sucking bug species occurred in all regions and on most crops, yet others were observed at only one or two sites. Green gram and cowpea were usually host to an abundance of sucking bugs, but adjacent areas of black gram sustained only sparse populations.

Most of the pest species reported are not currently considered to be major pests of grain legumes. However, their feeding activities make them worthy of mention. Further, many of these "minor" pests may act together in sufficient numbers to cause significant damage to the crop. This was noted in the Ord River Irrigation Area, where at least six species of pod-sucking insects occurred together and caused severe damage to cowpeas and mungbeans.

A number of additional potentially important pest groups, such as whiteflies, mites, jassids, aphids, wireworms, termites, and cutworms, are not comprehensively covered. These groups were not observed in high densities in any of the areas surveyed.

The most abundant potentially beneficial arthropods in terms of species recorded were the wasps (Hymenoptera; 39 per cent), beetles (Coleoptera; 13 per cent), flies (Diptera; 10 per cent), bugs (Hemiptera; 9 per cent), and the spiders (Araneida; 21 per cent). In this booklet, the beneficial species are classified as either predators or parasites. Predacious species are those which are free-living and attack and consume the eggs, larvae, or adults of others.

In total, over a thousand predatory arthropod species have been identified from soybean fields in the United States, and it is likely that the predator complex will prove to be equally diverse in Australia. The predaceous species reported here therefore represent only a portion of the total. Grossly under-represented, for example, are the ants (Formicidae) which can play a significant predaceous role in grain legume crops.

Parasites are free-living as adults but in their juvenile stages develop in or on host insects. They are most valuable in suppressing certain pests, but their presence and abundance are often difficult to assess. There are no reliable procedures for sampling adult parasites in the field. The most practical approach for sampling parasites is to collect potential hosts from the field, rear them in the laboratory, and record the emergence of adult parasites. In this study there was only limited success in recording parasites because many of the insects collected, particularly various caterpillars, were unable to live on the artificial diets provided. The limited number of parasite species in this report should therefore not be taken as a true indication of their importance. In South Carolina, for example, some thirty-three parasite species are reported to attack soybean caterpillar pests alone (Sanders 1978).

Included in the beneficial organisms in this booklet are several insect diseases which can play a major role in controlling pest outbreaks. For example, the fungus *Nomuraea rileyi* is credited with being the major controlling factor in outbreaks of several caterpillar pests in soybeans in the south-eastern United States (Carner, Shepard, and Turnipseed 1974*a*).

PESTS

Plant-Sucking Bugs

Insects that pierce the plant and feed on its juices form the most abundant complex of insect pests of grain legumes. Many of them are pod suckers and cause substantial yield loss by piercing the pod wall and feeding directly from the developing seeds. However, assessment of losses caused by pod-sucking insects can be more difficult than with the pod-chewing insects, because the damage may not be readily visible. The degree of damage depends on the stage of development of the pod and seed. When flowers and young pods are attacked, they may fall from the plant. Damaged ovules will usually shrivel and die (fig. 12). However, older seeds may show only puncture marks (Miner 1966, Turner 1977), and damage may not be detected until after harvest. Feeding by pod-sucking insects may cause a reduction in seed quality and germinability and in soybean may lower oil content. Causal agents for diseases may be transmitted to the seed (Daugherty et al. 1964), or the seed may be made vulnerable to attack by diseases during subsequent germination (Lawn and Russell 1978).

The green vegetable bug, *Nezara viridula* (fig. 24), was considered the major pest of grain legumes in many areas (Williamson 1971, Lawn and Russell 1978, Passlow and Waite 1971). However, its importance in some areas has decreased in recent years because of the introduction and establishment of the egg parasite *Trissolcus basalis* (fig. 99). The economic control of the green vegetable bug in the Ord River Irrigation Area is attributed to the successful establishment of this parasite.* Chemical control is still necessary in areas where *T. basalis* is not well established.

Eggs of *N. viridula* are laid in rafts on the underside of leaves, and

* G. Strickland 1979: personal communication

nymphs emerge in approximately seven days depending upon temperature. All five nymphal stages and adults feed on developing pods (fig. 25), a pattern generally true of other pod-sucking species. Other bugs that may cause economic damage as individual species include the pod-sucking bugs *Riptortus serripes* (fig. 14) and *R. linearis;* the brown bean bug, *Melanacanthus scutellaris* (fig. 13); and the redbanded shield bug, *Piezodorus hybneri* (fig. 27). Generally most damage arises from two or more species occurring together. For example, it was not uncommon to find numbers of *R. serripes* and *M. scutellaris* together at economically damaging levels in all regions surveyed.

In the Ord River Irrigation Area, cowpeas and green gram supported damaging communities of a range of pod-sucking bugs, including *P. hybneri; R. serripes; M. scutellaris;* the paddy bug, *Leptocorisa acuta; Acroelytrum muricatum* (fig. 15); and the crusader bug, *Mictis profana* (fig. 17). The passionvine bug, *Leptoglossus australis* (fig. 16), was found in large numbers on cowpea at Katherine, while sizeable populations of the brown shield bug, *Dictyotus caenosus* (fig. 22), were observed causing damage to developing seeds in soybean on the Darling Downs.

Other pod- and plant-sucking species which were almost universally present included the lygaeid bugs *Graptostethus servus* (fig. 18), *Melanerythrus mactans* (fig. 19), and the Rutherglen bug, *Nysius vinitor.* Large populations of *M. mactans* have damaged experimental green gram crops in south-eastern Queensland. The green mirid, *Creontiades dilutus* (fig. 20), and several pentatomid bugs — the horehound bug, *Agonoscelis rutila* (fig. 21); *Eysarcoris trimaculatus* (fig. 23); *Oncocoris* spp. (fig. 26); and the green stink bug, *Plautia affinis* (fig. 28) — were found at several locations. A number of minor sucking bugs species were also found (see index 1), but their populations did not reach economic densities in the areas surveyed.

Pod-Chewing Insects

While many of the pod-chewing insects feed also on foliage, blooms, and sometimes petioles and stems, most of the economic loss arises from direct feeding on the pods. Both seed and pod may be consumed, and damage to the pod wall may expose uneaten seed to weathering and to germination during rainy periods (figs. 56, 63). Although there are many species that have been observed to chew into pods, there are only relatively few species that currently cause serious economic loss. These include the cotton bollworm, *Heliothis armiger* (fig. 40), perhaps the most serious pest in Queensland; the native budworm, *Heliothis punctiger* (fig. 41); the bean podborer, *Maruca testulalis* (figs. 54, 55); and the lucerne seed web moth, *Etiella behrii* (fig. 51).

Heliothis spp. were found in all locations and crops, and in many cases constituted the most serious pest problems. Species within the genus *Heliothis* are probably the most important economic pests on a world-wide basis (Kogan et al. 1978), so it is not surprising that most grain legumes are favourable hosts. Larvae of *Heliothis* spp. feed on leaves, stems, flowers, and fruits of a wide range of cultivated and wild hosts as well as grain legumes.

Eggs and larvae of *H. armiger* and *H. punctiger* are similar and difficult to distinguish in the field. The colour of the larvae is variable (see figs. 42, 76, 83, 88, 89) and cannot be used for identification. Adults of *Heliothis* are more easily distinguished. *H. armiger* has a faint white mark on the dark band of each hind wing (fig. 40), which is absent on hind wings of *H. punctiger* (fig. 41). Colouration and markings on the forewings of these species are too variable to be of value in separating the species.

The host range, life history, description, control, and distribution of the bean podborer, *Maruca testulalis,* have been reported by Broadley (1977). Like *Heliothis,* the bean podborer feeds on leaves, stems, flowers, and pods of many grain legumes. It has been reported from eastern coastal and sub-coastal regions from central New South Wales to Cape York, and in this survey it was found in every area where grain legumes were grown.

The bean podborer is particularly destructive to mungbeans, cowpeas, and pigeonpeas, where the pods are arranged in such a manner as to allow the pest to web the pods and adjacent leaves together to form a protective shelter (fig. 54). In green gram fields in Darwin and Katherine, populations of the bean podborer and a lesser-known pod-feeding caterpillar, *Eublemma dimidialis* (figs. 38, 39), occurred together.

Caterpillars of the lucerne seed web moth, *Etiella behrii* (fig. 51), were widely observed causing damage to soybean pods in south-eastern Queensland. Serious outbreaks of *Etiella* sp. have been observed in Brazil.*

---

* M. Shepard 1980: personal observation

Caterpillars of various lycaenid butterflies, such as the bean flower caterpillar, *Jamides phaseli* (fig. 31); the pea blue butterfly, *Lampides boeticus;* the grass blue butterfly, *Zizina labradus;* and *Catochrysops panormus platissa* were present in most areas, particularly on black gram and pigeonpeas. A heavy infestation of *Z. labradus* larvae (fig. 32) was observed webbing leaves, pods, and flowers together on pigeonpeas at Katherine. *Euchrysops cnejus cnidus* (Lycaenidae) has also been reported as a pest of grain legumes (Radunz 1979). Another caterpillar, *Gymnoscelis lophopus* (fig. 30), has caused considerable flower damage in experimental areas of green gram in the Burnett region.

Locusts were occasionally observed in grain legumes, and pod damage from their feeding was often apparent (fig. 63). The spur-throated locust, *Austracris guttulosa* (fig. 57), and the Australian plague locust, *Chortoicetes terminifera* (fig. 59), appeared most abundant, but others such as the giant grasshopper, *Valanga irregularis* (fig. 60), the yellow-winged locust, *Gastrimargus musicus,* and the migratory locust, *Locusta migratoria,* were widespread.

Larvae of a small fly, *Melanagromyza pseudograta* (fig. 9), were recovered from soybean pods at sites throughout south-eastern Queensland. The fly develops entirely within the pod, pupation occurring within the empty locule created by the feeding larva (fig. 10). It was previously reported on the native perennial glycine, *Glycine tomentosa* (Spencer 1977), and is widely distributed throughout eastern and northern Australia. Another agromyzid podfly with similar feeding behaviour, *M. obtusa* is a major pest of pigeonpeas in India (ICRISAT 1977).

Foliage-Feeding Insects

Most of the species that chew into the pods of grain legumes also feed to some extent on foliage. As mentioned earlier, some of the caterpillar species web leaves, flowers, and pods together, causing damage to all of these plant parts. However, grain legumes can tolerate considerable damage to the foliage without yield being affected, and few fields surveyed were considered to be damaged above economic thresholds by foliage feeders alone.

For the most part, the moth/butterfly complex contains the largest number of important foliage-feeding species, with the Noctuidae by far the most economically important family. Several fields contained relatively high numbers of the tobacco looper, *Chrysodeixis argentifera* (fig. 35), and the green looper, *Chrysodeixis eriosoma* (figs. 36, 37). The cluster caterpillar, *Spodoptera litura* (figs. 49, 50), was especially common in the Lockyer Valley and in the Ord River Irrigation Area, but distribution of these caterpillars was clumped and at sub-economic levels. Armyworms in the same genus (*S. frugiperda* and *S. exigua*) sometimes occur in outbreak proportions in soybeans in the south-eastern United States. The northern armyworm, *Mythimna separata* (figs. 45, 46), was found in peanuts in the Ord.

Other loopers which were widely encountered included the castoroil looper, *Achaea janata* (figs. 33, 34); the cotton looper, *Anomis flava; Mocis alterna* (figs. 43, 44); and *Saroba irrorata* (figs. 47, 48). The soybean looper, *Diachrysia orichalcea,* was found in the South Burnett region where it has occasionally caused economic damage to soybean crops.* The beet webworm, *Hymenia recurvalis* (figs. 52, 53), was common on crops in all regions. In contrast, other species such as the heliotrope moth, *Utetheisa pulchelloides* (fig. 29), were of generally isolated occurrence.

A number of beetle species were observed feeding on the foliage of grain legumes. Both the plain pumpkin beetle, *Aulacophora abdominalis* (fig. 4), and the pumpkin beetle, *A. hilaris,* were common in all areas, although they were not believed to cause appreciable damage. The redshouldered leaf beetle, *Monolepta australis* (fig. 6), was observed in high numbers in one corner of a soybean field in the Lockyer Valley, but observation revealed that the density of this population declined after a week or two. The twentyeightspotted ladybird, *Henosepilachna vigintioctopunctata* (fig. 7), was also common, especially on mungbeans. A similar foliage-feeding beetle, the Mexican bean beetle *Epilachna varivestis,* is an important pest of soybeans and *Phaseolus* spp. in several parts of North America. Other leaf-eating beetles occasionally recorded were chrysomelid beetles such as *Cryptocephalus parentheticus* (fig. 5), and weevils such as the whitefringed weevil, *Graphognathus leucoloma* (fig. 8).

Other foliage feeders included locusts (see also under "Pod-Chewing Insects") such as *Bermius odontocercus* (fig. 58), grasshoppers such as *Attractomorpha similis* (fig. 62), and crickets. One leaf-eating cricket, *Madasumma planiceps* (fig. 61), was quite common in the Ord River Irrigation area. However, populations were not considered economically important. As

---

* J. Rogers and H. Brier 1979: personal communication

well as the various insect groups, spider mites may occasionally cause damage to foliage in localized areas (fig 64), especially where the crop has a history of high insecticide use.

Stem-Borers and Root-Feeding Insects

The lucerne crownborer *Zygrita diva* (fig. 1) is a major pest of soybeans (Wood 1976). Its host range includes lucerne and the native legumes *Sesbania* sp. and *Crotalaria* sp. In this survey, a few larvae were also found in pigeonpeas, guar, and green gram. Adult females prepare egg-laying sites by chewing small areas of stem or petiole. Larvae (fig. 2) and pupae (fig. 3) develop inside the plant, and overwintering occurs in the plant stem just below ground level. Several generations are possible in a soybean crop.* Infested plants are susceptible to lodging and in many cases mature prematurely when the lower stem is girdled by the larva before pupation. Control with insecticides appears unpractical, although chemical control has been achieved in experimental plots (Wood 1976). Control of a similar soybean stem-borer, *Dectes texanus texanus* has been achieved in the United States by ploughing in the crop stubble immediately after harvest (Campbell and Van Duyn 1977). Other stem-borers recorded from soybean in the Lockyer Valley were the speckled longicorn, *Paradisterna plumifera*, and *Corrhenes* spp. Stem-boring by larvae of *Etiella behrii* was also observed in soybeans.

The bean fly, *Ophiomyia phaseoli* (fig. 11), was widespread, and damage from tunnelling larvae was recorded on soybeans, black gram, green gram, adzuki beans, cowpeas, and phasey beans. Most frequently damage was isolated and occurred in out-of-season crops. Seedlings appeared to be most susceptible, and in one instance an experimental sowing of cowpeas was destroyed. A related species, the soybean fly, *Melanagromyza sojae*, was recovered from soybean stems in the Lockyer Valley. Both species are reported as major pests of soybeans, black gram, and green gram in south-east Asia (AVRDC 1979).

Larvae of several beetle species feed on roots of grain legume plants. Larvae of *Amnemus superciliaris* weevils were observed causing severe damage in one soybean field on the Darling Downs, and occasional

* G. Strickland 1979: personal communication

damage by false wireworm *Gonocephalum* sp. was noted. Adults of a number of tenebrionid and curculionid species were also found, but their precise role was not determined.

BENEFICIAL ORGANISMS

Predators

The relative abundance of certain species of predators in these surveys changed not only on a seasonal basis but also with different geographical locations and occasionally with different crops. The predator complex includes a diverse range of groups such as beetles, ants, wasps, spiders, assassin bugs, stink bugs, earwigs, flies, mantids, lacewings, dragonflies, and damselflies. There are usually fewer than a dozen species, however, which are abundant enough to make a significant impact by themselves on pest populations. The significance of the individual predacious species thus lies in their role as part of the overall beneficial complex. For example, the labelling studies with *Heliothis* spp. in soybeans indicated that, collectively, predacious species can substantially reduce populations of these pest species. The following examples of naturally occurring predators are not necessarily in order of their importance.

Some of the most abundant and conspicuous predators in crop canopies were the ladybirds, which were found in all the areas surveyed. The most frequently encountered species were the transverse ladybird, *Coccinella repanda* (fig. 70); the common spotted ladybird, *Harmonia conformis* (fig. 71); the minute twospotted ladybird, *Diomus notescens; Harmonia arcuata;* and *Micraspis aphidectoides.* These beetles, along with the pollen beetle, *Dicranolaius bellulus* (fig. 73), actively search the foliage for eggs and young larvae of most caterpillar pests, and aphids and mites. Both adults and larvae (fig. 72) are predatory.

The soldier beetle *Chauliognathus pulchellus* (fig. 65) was encountered in large numbers at several sites in southern Queensland. A related species, *C. nobilitatus,* was often found in small numbers in association with *C. pulchellus.* Larvae and adults of these species are predacious, feeding on soft-bodied insects.

In many locations ants were observed actively foraging over plants, particularly in cowpea and mungbean crops. Most frequently encountered were meat ants *Iridomyrmex* sp. (fig. 94), which were observed attacking live larvae of *Maruca testulalis* and *Spodoptera* sp. Usually when *Iridomyrmex* was abundant, plants were almost devoid of caterpillar pests.

Important foliage-dwelling predatory bugs which were capable of attacking and consuming larger caterpillar larvae included the predatory shield bug *Oechalia schellenbergii* (fig. 80), and the reduviid bugs *Coranus triabeatus* and *Sastrapada australica.* Nymphs of these bugs are also predatory (fig. 81). The largest predatory bug, the assassin bug, *Pristhesancus papuensis* (fig. 83), can attack and consume the largest *Heliothis* larva. The long life span and fecundity of adult females together with the predatory nature of all nymphal stages make this beneficial insect worthy of consideration for use in a biological control programme (McWhorter and Shepard 1982).

Other predacious bugs found were the shield bug *Cermatulus nasalis* (fig. 79) and the assassin bug *Gminatus nigroscutellatus* (fig. 82). A damsel bug, *Tropiconabis nigrolineatus,* was an abundant predator in most soybean fields and adjacent vegetation in south-eastern Queensland. Other relatively abundant and potentially important bug species included the apple dimpling bug, *Campylomma livida,* and the bigeyed bug, *Geocoris lubra.* Although these species feed to some extent on plant tissue, they were observed feeding on eggs and young larvae of several pest species.

Besides the rich complex of predators that reside in the canopy of grain legumes, there are many species that fly into the fields in search of prey but complete their life cycles elsewhere. Because of this behaviour, it is difficult to sample these species and accurately assess their numbers. They include the dragonflies, such as *Diplacodes bipunctata* (fig. 102), and the damselflies, e.g., *Ischnura aurora* (fig. 103). Predacious wasps were prevalent in some crops, particularly where there were woodlands adjacent to the field. Solitary wasps found foraging within crops included *Anterhynchium nigrocinctum nigrocinctum* (fig. 91), *Epiodynerus tamarinus* (fig. 92), *Pseudepipona bicolor* (fig. 93), *Ferreola handschini* (fig. 98), and *Priononyx saevus* (fig. 100). Wasps that seek out caterpillars to carry back to communal nests included *Polistes variabilis* and *Ropalidia gregaria.* In several areas, robber flies such as the wasp mimic *Chrysopogon* sp. (fig. 75) were found hunting within the crop canopy.

Some of the most important and least understood of the predatory insects in grain legumes are the ground-dwelling beetles and earwigs.

These predators forage at night and can easily be overlooked. Carab beetles such as *Chlaenius flaviguttatus* (fig. 67), *Gnathaphanus pulcher* (fig. 68), and the green carab beetle, *Calosoma schayeri* (fig. 66), can often be found by looking under the debris at the base of plants. Carab beetle larvae (fig. 69) are also important predators within the canopy. The common brown earwig, *Labidura riparia truncata* (fig. 74), was one of the most abundant and important predators of *Heliothis* spp. eggs and young larvae in radio-isotope experiments in south-eastern Queensland.

Other predators occasionally found included mantids such as *Orthodera ministralis* (fig. 104) and adult lacewings such as *Chrysopa* sp. (fig. 101). Lacewing larvae camouflaged with small pieces of plant debris attached to their bodies were frequently observed feeding on aphids.

Spiders are one of the most abundant groups of predators in grain legumes. Besides being obligate predators feeding on both pest and beneficial species, they are an important source of food for other predatory species (Whitcomb 1980). Although spiders were not comprehensively sampled during these surveys, some of the more frequently encountered species were collected from the plant canopy. The importance of many ground-dwelling species was recognized, but assessment of their presence and abundance would have required intensive sampling using pitfall traps.

Spiders catch their prey in several different ways. Some spiders use webs which are spun within the canopy to catch flying insects; examples of these are *Araneus heroine* (fig. 105), *Latrodectus mactans hasselti* (fig. 111), and *A. theisi.* Others either stalk their prey or lie in wait within the canopy. The yellow nightstalker, *Chiracanthium diversum* (fig. 106), spends the daylight hours in a webbed chamber on the undersides of leaves. The wolf spiders, *Lycosa* spp. (fig. 107), normally hide under debris or in a burrow or cracks in the soil during the day but at night climb into the canopy in search of prey.

The lynx spiders — for example, *Oxyopes macilentus* (fig. 108) and *O. molarius* (fig. 109) — are active in the daytime and were some of the most frequently encountered spiders in areas surveyed. Jumping spiders such as *Cytaea* sp. (fig. 110) patrol the foliage in search of prey and have been reported to feed on insect eggs. Crab spiders such as *Diaea* sp. (fig. 112) usually lie in wait for their prey and are often seen feeding upon insects several times their size.

The full significance of spiders as beneficial arthropods in agricultural ecosystems is just beginning to be recognized by those involved in conservation of natural enemies in grain legumes. As more is understood about this important group of predators, perhaps they can be better conserved and encouraged.

Parasites

Parasites are extremely valuable in maintaining certain pests below the damage threshold. In general, parasitic species are much more susceptible to chemical insecticides than are predators, and the rapid increase in pest density after an insecticide application is at least partially a result of the destruction of naturally occurring parasites.

Among the important naturally occurring parasites, the braconid wasps were represented by at least twelve species which emerged from pest caterpillars. *Microplitis demolitor* (figs. 87-89) was commonly found in *Heliothis* spp. larvae. Two *Bracon* spp. parasites (fig. 85) of *Maruca testulalis* were recorded from green gram fields, one at Katherine and the other in the Burnett region. *Apanteles* spp. (fig. 84) were common, particularly on *Chrysodeixis* spp. larvae. *Cremnops dissimilis* (fig. 86) was frequently observed flying within crop canopies.

The encyrtid wasp *Litomastix* sp. was common in populations of the tobacco looper, *Chrysodeixis argentifera,* in soybeans in the Lockyer Valley (Qld.). This tiny parasite lays its eggs in the eggs of its host and later develops within the host caterpillar. By the time the caterpillar is mature, the original single parasite egg has divided to produce around a thousand parasites which ultimately emerge from the dead and mummified looper (fig. 90).

The orange caterpillar parasite, *Netelia producta* (fig. 97), a parasite of *Heliothis* spp., can occur in considerable numbers. In south-eastern Queensland adult wasps were frequently observed flying around the canopy of mungbean and soybean crops in the mid to late afternoon, when caterpillars of *Heliothis* were observed to emerge from the canopy to feed on exposed pods and foliage. On several occasions adult female wasps were observed attempting to parasitize exposed caterpillars. Two other ichneumonid wasps — the twotoned caterpillar parasite, *Heteropelma scaposum* (fig. 95), and the orchid dupe, *Lissopimpla excelsa* (fig. 96) — are also common parasites of caterpillars. In one soybean field at Coominya in the West Moreton region, parasitism by *H.*

scaposum and *M. demolitor* collectively resulted in 20 per cent mortality of some four hundred *Heliothis* larvae collected and reared.

A number of parasitic fly species were recovered from caterpillar and grasshopper pests. Tachinid fly eggs were frequently observed attached to *Heliothis* larvae (fig. 76), and adult flies (fig. 77) emerged from *Heliothis* larvae collected at all sites surveyed. Tachinids were also recovered frequently from larvae of *Spodoptera litura* and *Chrysodeixis* spp. Adults of two locust species, *Austracris guttulosa* and *Valanga irregularis,* were parasitized by tachinid fly larvae (fig. 78).

Insect Diseases

The cursory nature of the surveys did not allow close monitoring of disease outbreaks. However, occasional disease epizootics were noticed in the course of the surveys. An outbreak of the fungus *Nomuraea rileyi* (fig. 114), for example, completely controlled an economically important population of the tobacco looper, *Chrysodeixis argentifera,* in soybeans at Forest Hill. Virus-infected larvae (fig. 115) were frequently observed in populations of *Heliothis* spp., *Chrysodeixis* spp., *Spodoptera litura,* and *Anomis flava.* During one rainy period, approximately half the population of *Heliothis* in a soybean field were killed by virus. Occasionally the fungus *Entomophthora* sp. (fig. 113) was observed attacking loopers.

There are several distinct advantages associated with the use of "microbial agents" in integrated control programmes, which are discussed by Teakle (1977). Whether the disease organism is a fungus, a virus, or some other microbe, it is normally self-perpetuating and without the harmful side effects of chemical insecticides. Generally, the well-developed canopy of grain legumes provides a favourable micro-habitat for the development of beneficial disease organisms, especially fungi. Recent research has shown, however, that application of fungicides for control of plant pathogens in soybeans can suppress *N. rileyi* and result in outbreaks of foliage-feeding caterpillars. The potentially important viruses of *Heliothis* have been described by Teakle (1973, 1974, 1977).

Fig. 1. Adult of the lucerne crownborer *Zygrita diva* feeding on the native legume *Sesbania cannabina,* one of its alternative hosts (Ord River). Colour is variable, usually brown, fawn, grey, or black. The dark markings are quite variable in size and usually number (in pairs) 0–6. Length 8–12 mm. Another lucerne crownborer, *Corrhenes stigmatica,* is generally smaller and is distinguished by more mottled grey-brown colouring.

Fig. 2. *Zygrita diva* larva tunnelling in soybean stem (Lockyer Valley).

Fig. 3. Soybean stem split to expose *Zygrita diva* pupa (Lockyer Valley).

Fig. 4. Adult of the plain pumpkin beetle, *Aulacophora abdominalis*, on cowpea (Burnett). Its plain colour distinguishes this beetle from the pumpkin beetle, *A. hilaris*, which has four dark markings on its wing covers. Length 7 mm.

Fig. 5. *Cryptocephalus parentheticus* adult showing its characteristic wing-cover markings, on soybean (Lockyer Valley). Length 5 mm.

Fig. 6. Adult of the monolepta beetle or redshouldered leaf beetle, *Monolepta australis,* showing characteristic red markings at the base of its wing covers, on soybean (Lockyer Valley). Length 5 mm.

Fig. 7. Twentyeightspotted ladybird, *Henosepilachna vigintioctopunctata,* adult on soybean (Ord River). Despite the name, the number of spots varies. Length 6 mm.

Fig. 8. Adult of the whitefringed weevil, *Graphognathus leucoloma,* feeding on a soybean leaflet (South Burnett). This weevil is distinguished by the white band along the side of its wing cover. Approximate length 10 mm.

Fig. 9. Adult of a pod fly, *Melanagromyza pseudograta,* on a soybean pod (Darling Downs). The exit hole from which it has emerged is to the left. This species is slightly larger than the bean fly (fig. 11).

Fig. 10. Soybean pod split to expose a larva of *Melanagromyza pseudograta* feeding on a developing seed (Darling Downs).

Fig. 11. Adult bean fly, *Ophiomyia phaseoli,* on cowpea leaflet exhibiting characteristic damage symptoms from egg-laying activity (Darwin).

Fig. 12. Example of sucking-bug damage in cowpea. Stung seeds are characteristically shrivelled with discoloured necrotic areas. Undamaged seed is on the right.

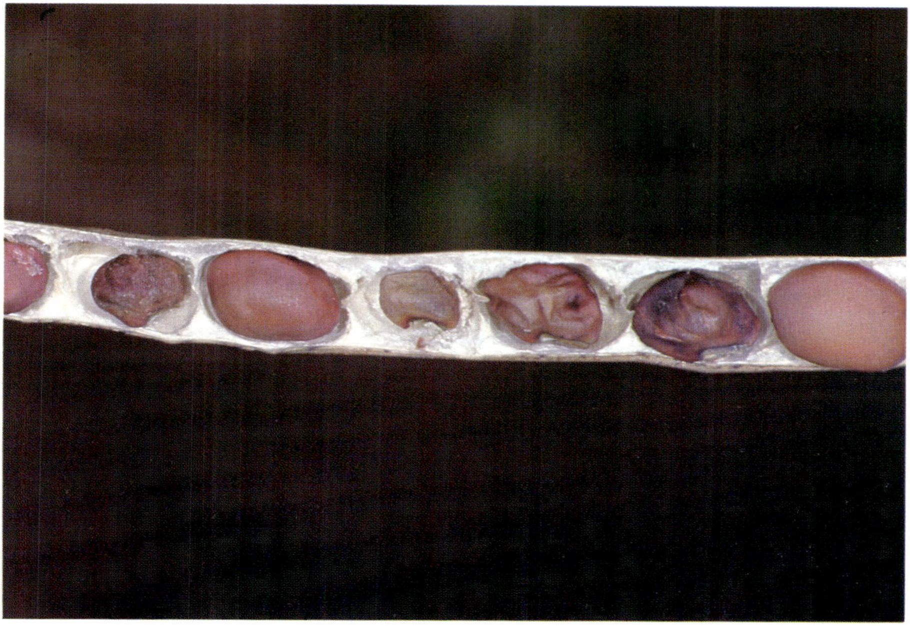

Fig. 13. Adult of the brown bean bug, *Melanacanthus scutellaris,* on green gram (Ord River). This bug is smaller than the podsucking bug (fig. 14) and is distinguished from it by the pale yellow patch between the wing bases on its back. Length 10–12 mm.

Fig. 14. Adults of the podsucking bug, *Riptortus serripes,* on cowpea (Katherine). This bug is characterized by a pale yellow band along the side of its body. Length 14–18 mm.

Fig. 15. *Acroelytrum muricatum* adult on soybean (Ord River). Approximate length 17 mm.

Fig. 16. Adult of the passionvine bug, *Leptoglossus australis,* on cowpea (Katherine). This bug is readily recognized because of the presence of bright orange spots on its underside. Length 17–19 mm.

Fig. 17. Adult crusader bug, *Mictis profana*, showing the characteristic light-coloured cross on its back (Ord river). Length 21–23 mm. Note the proboscis piercing the cowpea pod. The long, slender, straight proboscis is typical of the plant-sucking bugs and contrasts with that of the predatory bugs (e.g., figs. 82, 83).

Fig. 18. Plant-sucking bug *Graptostethus servus* adult on soybean leaflet (Ord River). Length 7–8 mm.

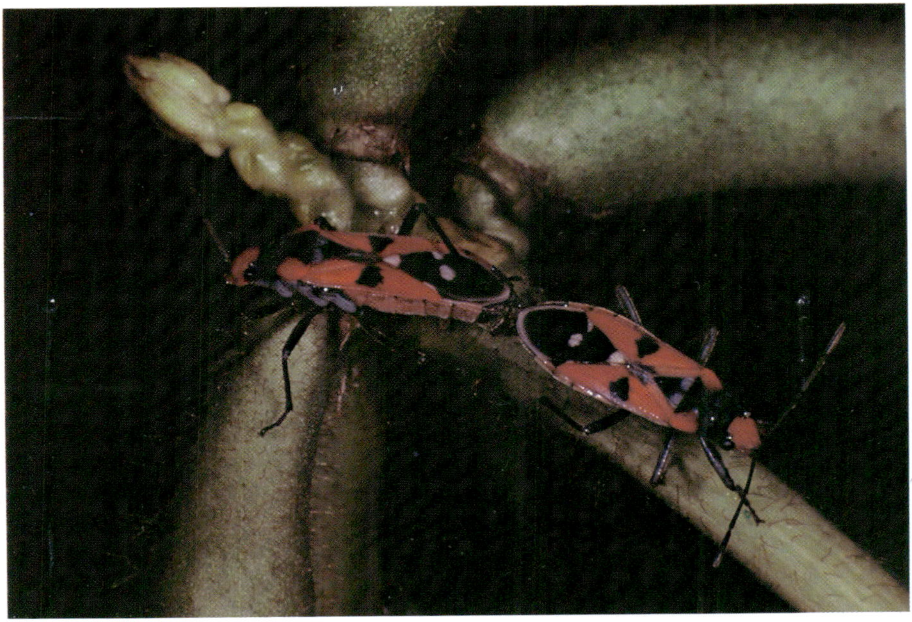

Fig. 19. Adults of *Melanerythrus mactans,* clearly showing the very characteristic markings, on green gram (Ord River). Length 8–10 mm.

Fig. 20. Adult of the green mirid, *Creontiades dilutus,* on green gram (Ord River). Length 5–6 mm.

Fig. 21. Adult horehound bug, *Agonoscelis rutila,* showing characteristic markings, on green gram pods (Darling Downs). Length 10 mm.

Fig. 22. Adult of the brown shield bug, *Dictyotus caenosus,* on soybean (Darling Downs). Length 7–10 mm.

Fig. 23. Adult of *Eysarcoris trimaculatus* on cowpea (Katherine). It is distinguished by the two pale spots near the bases of the wings. Length 5–6 mm.

Fig. 24. Adult of the green vegetable bug, *Nezara viridula,* feeding on soybean pod (Lockyer Valley). Length 14–18 mm.

Fig. 25. Nymphs of the green vegetable bug, *Nezara viridula,* clustered on green gram pods (Darwin).

Fig. 26. Adults of *Oncocoris hackeri* on cowpea pod (Ord River). This bug is distinguished from the brown shield bug (fig. 22) by the darker, more mottled colouring. Length 7–9 mm.

Fig. 27. Adult of the redbanded shield bug, *Piezodorus hybneri,* on soybean (Ord River). It is distinguished by the band (usually red or pink, occasionally white) across the thorax, and it is paler green and smaller than the green vegetable bug (fig. 24). Length 9–10 mm.

Fig. 28. Adult of the green stink bug, *Plautia affinis,* showing characteristic black wing tips, on soybean pod (Lockyer Valley). Length 8 mm.

Fig. 29. Adult heliotrope moth, *Utetheisa pulchelloides,* in soybean (Ord River). Length 14–16 mm.

Fig. 30. *Gymnoscelis lophopus* larva on a green gram inflorescence (Burnett).

Fig. 31. Adult of the bean flower caterpillar, *Jamides phaseli* (Lockyer Valley). Wingspan 25 mm.

Fig. 32. Larva of the grass blue butterfly, *Zizina labradus,* on a pigeonpea flower (Katherine).

Fig. 33. Castoroil looper, *Achaea janata,* on cowpea (Ord River). Colour is variable, frequently blue-black in older caterpillars.

Fig. 34. Moth of the castoroil looper, *Achaea janata* (Burnett). Wingspan 50–55 mm.

Fig. 35. Moth of the tobacco looper, *Chrysodeixis argentifera* (Lockyer Valley). It is distinguished from the adult of the green looper (fig. 36) by the oblique silver S midway between the large silver spot and front margin of the forewing. Wingspan 30–35 mm.

Fig. 36. Moth of the green looper, *Chrysodeixis eriosoma*, on cowpea (Katherine).

Fig. 37. Green looper, *Chrysodeixis eriosoma*, on soybean (Lockyer Valley). *Chrysodeixis* spp. caterpillars are distinguished by the white stripes, green colour, and the presence of only two pairs of abdominal prolegs.

Fig. 38. Larva of *Eublemma dimidialis* feeding on a green gram pod (Darwin).

Fig. 39. Moth of *Eublemma dimidialis* showing characteristic colourful markings (Darwin). Wingspan 18–20 mm.

Fig. 40. Adult of the cotton bollworm, *Heliothis armiger*, displaying characteristic white marking in the dark fringe of the left hind wing (Lockyer Valley).

Fig. 41. Adult of the native budworm, *Heliothis punctiger* (Lockyer Valley). It is distinguished from the adult of the cotton bollworm (fig. 40) by the absence of a white spot in the dark band of the hind wing.

Fig. 42. *Heliothis* sp. larva on green gram (Burnett). Examples of other colour morphs can be seen in figs. 76, 83, 88, 89. *Heliothis* spp. caterpillars have four pairs of prolegs, and the skin is rough to touch because of body hairs on the prominent tubercles.

Fig. 43. Adult of *Mocis alterna* in soybean showing characteristic forewing pattern and pale and dark bands on the hind wing (Lockyer Valley). Wingspan 33–36 mm.

Fig. 44. Larva of *Mocis alterna* on soybean (Ord River).

Fig. 45. Larva of the northern armyworm, *Mythimna separata,* on peanut (Ord River).

Fig. 46. Moth of the northern armyworm, *Mythimna separata* (Ord River). It is distinguished from *Heliothis* spp. adults by the absence of a distinct dark band on the hind wing (not shown). Wingspan 38–40 mm.

Fig. 47. Larvae of *Saroba irrorata* feeding on a green gram leaflet (Lockyer Valley). They are distinguished from *Chrysodeixis* spp. caterpillars (fig. 37) by the very distinct band along the side and the four pairs of abdominal prolegs (see lower specimen).

Fig. 48. Moth of *Saroba irrorata* in green gram showing characteristic oblique line across the wings and B-shaped spot on the forewings (Darwin). Wingspan 35–40 mm.

Fig. 49. The cluster caterpillar, *Spodoptera litura,* on soybean (Lockyer Valley). It is distinguished from *Heliothis* spp. caterpillars by the smooth skin and the three light-coloured stripes on top of the first segment behind the head.

Fig. 50. Moth of the cluster caterpillar, *Spodoptera litura,* showing the characteristic brown and white markings on the forewings (Lockyer Valley). Wingspan 30–45mm.

Fig. 51. Larva of the lucerne seed web moth, *Etiella behrii,* on a soybean pod in which it had been feeding (Darling Downs).

Fig. 52. Beet webworm, *Hymenia recurvalis,* exposed within a folded cowpea leaflet (Burnett).

Fig. 53. Moth of the beet webworm, *Hymenia recurvalis,* showing characteristic dark brown and white colouring, on green gram (Burnett). Wingspan 20–22 mm.

Fig. 54. A bean podborer, *Maruca testulalis,* dislodged from its shelter within a green gram pod (Ord River). *M. testulalis* is usually readily distinguished from the larvae of *Eublemma dimidialis* (fig. 38) and of the lucerne seed web moth (fig. 51) by the presence of the paired, dark plates on top of the thoracic segments (see also fig. 85).

Fig. 55. Characteristic pose of the moth of the bean podborer, *Maruca testulalis* (Darwin). Wingspan 24–28 mm.

Fig. 56. Economic loss from podborers usually includes secondary loss from weathering damage. This green gram pod shows how the damage to the pod allows entry of water during rainy periods. The seed near the hole in the pod has already sprouted, and the three adjacent seeds have imbibed water and swollen, splitting the pod open.

Fig. 57. The spur-throated locust, *Austracris guttulosa*, on green gram (Ord River). This locust is distinguished from other species by the dark markings on the forewings and unmarked transparent hind wings (not shown). Length from head to wing tips 45–55 mm.

Fig. 58. *Bermius odontocercus* on lablab bean showing characteristic brown stripe on the side of the thorax (Burnett). Both sets of wings are unmarked. Length from head to wing tips 25–30 mm.

Fig. 59. Australian plague locust, *Chortoicetes terminifera* (Burnett). Adults may range in colour from green to brown and are distinguished from other species by black tips to otherwise transparent hind wings and by scarlet tibiae (shanks) on the hind legs (not shown). Length from head to wing tips 25–35 mm.

Fig. 60. Adult of the giant grasshopper, *Valanga irregularis*, on soybean (Ord River). There is a large range in body size, colour, and extent of markings on the forewings. Length from head to wing tips 50–85 mm.

Fig. 61. Adult of a leaf-eating cricket, *Madasumma planiceps,* feeding on soybean (Ord River). Note how the leaf tissue has been stripped from between the network of veins.

Fig. 62. Green colour of *Attractomorpha similis* adult closely matches that of the soybean leaf (Ord River).

Fig. 63. Characteristic chewing damage from grasshoppers on a cowpea pod. Apart from the direct loss from feeding, secondary losses result from weathering of undamaged seed in the pod.

Fig. 64. Characteristic leaf mottling arising from mite damage in soybeans (Redland Bay). Damage from mites is usually seen in localized areas of a field and is usually more prevalent in dry seasons.

Fig. 65. Soldier beetles *Chauliognathus pulchellus* on green gram (Darling Downs). *C. pulchellus* is distinguished from *C. nobilitatus* by the narrow yellow band on the hind margin of the thorax. The entire thorax of *C. nobilitatus* is orange. Length 10–14 mm.

Fig. 66. Green carab beetle, *Calosoma schayeri,* foraging beneath a soybean canopy (Lockyer Valley). Length 12–15 mm.

Fig. 67. *Chlaenius flaviguttatus* adult showing characteristic orange spots on wing covers (Lockyer Valley). Length 15 mm.

Fig. 68. *Gnathaphanus pulcher* adult (Darwin). It is distinguished from the green carab (fig. 66) by the metallic reddish wing covers. Length 17 mm.

Fig. 69. Carab larva feeding on a *Maruca testulalis* larva (Katherine).

Fig. 70. Adult transverse ladybird, *Coccinella repanda,* feeding on aphids on a cowpea pod (Burnett). *C. repanda* is distinguished from other ladybird species by the wavy transverse black band across the wing covers. Colour may vary from red to yellow. Length 44 mm.

Fig. 71. Adult of the common spotted ladybird, *Harmonia conformis,* (Lockyer Valley). It is distinguished from the twentyeightspotted ladybird (fig. 7) by the presence of fewer but larger spots. Length 5 mm.

Fig. 72. A ladybird larva feeding on aphids on a green gram flower.

Fig. 73. The pollen beetle, *Dicranolaius bellulus,* on soybean (Darling Downs). Length 4 mm.

Fig. 74. Common brown earwig, *Labidura riparia truncata,* feeding on a *Heliothis* larva (Lockyer Valley).

Fig. 75. A robber fly *Chrysopogon* sp. (Burnett). This species is a wasp mimic. Length 10 mm.

Fig. 76. *Heliothis* sp. larva parasitized by a tachinid fly. The small black egg of the fly is attached laterally between the second and third pair of legs of the caterpillar (Lockyer Valley).

Fig. 77. Parasitic tachinid fly *Palexorista* sp. (Ord River). Length 7 mm.

Fig. 78. Larva of a parasitic tachinid fly emerging through the soft integument behind the head of a giant grasshopper, *Valanga irregularis*, before pupation (Brisbane).

Fig. 79. *Cermatulus nasalis*, a predatory shield bug, feeding on a *Heliothis* larva (Lockyer Valley). Length 10–13 mm.

Fig. 80. The predatory shield bug *Oechalia schellenbergii* feeding on a tobacco looper, *Chrysodeixis argentifera* (Katherine). *O. schellenbergii* is distinguished from *Cermatulus nasalis* (fig. 79) by the spines on the side of the thorax and the more slender body. Length 10 mm.

Fig. 81. Nymph of a predatory shield bug feeding on a cluster caterpillar, *Spodoptera litura,* on a green gram flower (Ord River).

Fig. 82. *Gminatus nigroscutellatus*, an assassin bug, feeding on a *Heliothis* caterpillar (Ord River). Note the short, strong, curved proboscis typical of predatory bugs. Length 12 mm.

Fig. 83. The assassin bug, *Pristhesancus papuensis,* feeding on a *Heliothis* caterpillar (Lockyer Valley). Note the strong curved proboscis piercing the prey.

Fig. 84. Pupal cocoons of a small parasitic wasp *Apanteles* sp. on a tobacco looper, *Chrysodeixis argentifera* (Lockyer Valley).

Fig. 85. Larvae of a small parasitic wasp *Bracon* sp. emerging from a dead *Maruca testulalis* caterpillar before pupation. The second caterpillar was unparasitized (Katherine).

Fig. 86. Adult parasitic wasp *Cremnops dissimilis* resting on the underside of a black gram leaf (Burnett). Length 8 mm.

Fig. 87. Adult parasitic wasp *Microplitis demolitor* ovipositing into a *Heliothis* caterpillar (Lockyer Valley). Usually only one egg is laid per caterpillar. Length 3 mm.

Fig. 88. After feeding internally on its host, a *Microplitis demolitor* larva emerges from a *Heliothis* caterpillar to pupate.

Fig. 89. Pupal case of *Microplitis demolitor* beside an immobilized and dying *Heliothis* larva on a pigeonpea pod (Lockyer Valley).

Fig. 90. Mummified looper of *Chrysodeixis* sp. from which adults of the tiny parasitic wasp *Litomastix* sp. are emerging (Lockyer Valley).

Fig. 91. Predacious wasp *Anterhynchium nigrocinctum nigrocinctum* hunting within a black gram canopy (Katherine). Length 11 mm.

Fig. 92. Predacious wasp *Epiodynerus tamarinus* hunting a bean podborer, *Maruca testulalis*. The borer is hiding in a protective shelter provided by a spent flower and leaf webbed together in this pigeonpea inflorescence. The wasp has chewed open a hole at one end of the leaf and is in the process of chewing a second hole to expose the trapped grub (Lockyer Valley).

Fig. 93. Predacious wasp *Pseudepipona bicolor* stinging a *Maruca testulalis* caterpillar after forcing it from its webbed protection in a pigeonpea inflorescence. The immobilized caterpillar is then removed to the wasp's nest (Lockyer Valley).

Fig. 94. Meat ants *Iridomyrmex* sp. attacking a *Maruca testulalis* larva on the ground beneath a pigeonpea crop (Lockyer Valley).

Fig. 95. The twotoned caterpillar parasite, *Heteropelma scaposum,* resting on a soybean leaf (Lockyer Valley). It is distinguished from the orange caterpillar parasite (fig. 97) by black thorax and yellow fore and mid legs. Length 15–20 mm.

Fig. 96. A female orchid dupe, *Lissopimpla excelsa* (Ord River). It is characterized by black wings and white patches (not visible) along the sides of the black segments of the abdomen. Length 11–16 mm.

Fig 97. The orange caterpillar parasite, *Netelia producta* (Burnett). Length 15–22 mm.

Fig. 98. Predacious wasp *Ferreola handschini* (Ord River). Length 20 mm.

Fig. 99. The green vegetable bug egg parasite, *Trissolcus basalis,* parasitizing a *Nezara viridula* egg raft. The tiny wasp moves systematically over the raft to parasitize most if not all of the bug eggs.

Fig. 100. Large predacious wasp *Prionomyx saevus* (Ord River). Length 25 mm.

Fig. 101. Adult green lacewing *Chrysopa* sp. in soybean (Lockyer Valley). Length 10–15 mm.

Fig. 102. A dragonfly *Diplacodes bipunctata* feeding on a plant-sucking bug *Euphanta* sp. in a ricebean crop (Lockyer Valley).

Fig. 103. A damselfly *Ischnura aurora* resting on a soybean leaf (Ord River).

Fig. 104. An example of another predator group, the mantids. The green mantid, *Orthodera ministralis*, clasps an adult of the lucerne crownborer *Zygrita diva* before feeding (Lockyer Valley).

Fig. 105. An orb-weaving spider *Araneus heroine* (Lockyer Valley).

Fig. 106. The yellow nightstalker, *Chiracanthium diversum*, is a common spider of grain legume crops (Lockyer Valley).

Fig. 107. A juvenile wolf spider *Lycosa* sp. well camouflaged on the ground beneath the crop canopy (Burnett).

Fig. 108. A successful lynx spider, *Oxyopes macilentus*, photographed with its catch in a black gram crop (Ord River).

Fig. 109. A lynx spider *Oxyopes molarius* feeding on a redbanded shield bug, *Piezodorus hybneri*, in soybean (Ord River).

Fig. 110. A jumping spider *Cytaea* sp. hunting in a soybean canopy (Ord River).

Fig. 111. In a grain legume crop, the redback spider, *Latrodectus mactans hasselti*, is a beneficial species, preying on a wide range of pest species (Katherine).

Fig. 112. A crab spider *Diaea* sp. feeding on an adult of the lucerne crownborer *Zygrita diva* in a soybean crop (Lockyer Valley).

Fig. 113. *Entomophthora* sp. fungus in the resting-spore stage on a mummified looper caterpillar.

Fig. 114. *Nomuraea rileyi* fungus enveloping a noctuid caterpillar just before spore production.

Fig. 115. A *Heliothis* caterpillar stricken with nuclear polyhedrosis virus. Characteristically the caterpillar hangs limply, remaining attached by its prolegs to the pigeonpea pod on which it was feeding.

POSTSCRIPT

Although the scope of the surveys and the information presented in this booklet are limited, the survey results clearly illustrate the diversity of potential pest species and their natural enemies found in grain legume crops in northern Australia. Of some significance is the fact that a large proportion of species could not be fully identified. In many cases these were previously unnoticed native species which have moved into the new environment created by the grain legume crop. The survey thus highlights the dearth of knowledge on the composition and biology of the potentially economic insects found in these areas and emphasizes the need for greater entomological research input in the development of stable cropping systems.

As cropping of grain legumes expands in the tropics and subtropics, the number of species of importance, both pest and beneficial, must be expected to increase. Nonetheless, the information presented should provide an appreciation of the current status of the insect pests and their natural enemies in the grain legumes. If the focus of this booklet on the beneficial as well as the pest species increases appreciation of the need for crop production systems that conserve and utilize natural regulating agents, it will have been particularly successful.

REFERENCES

AVRDC. 1979. Asian Vegetable Research and Development Center, Progress Report for 1978. Shanhua, Taiwan.

Broadley, R.H. 1977. The bean podborer in north Queensland. *Qld. Agric. J.* 103: 274-78.

Campbell, W.V., and Van Duyn, J.W. 1977. Cultural and chemical control of *Dectes texanus texanus* on soybeans. *J. Econ. Entomol.* 70: 256-58.

Carner, G.R.; Shepard, M.; and Turnipseed, S.G. 1974*a*. Disease incidence on Lepidopterous pests of soybeans. *J. Ga. Entomol. Soc.* 10: 101-5.

______. 1974*b*. Seasonal abundance of insect pests of soybeans. *J. Econ. Entomol.* 67: 487-93.

Daugherty, D.M.; Neustadt, M.H.; Gehrke, C.W.; Cavanah, L.E.; Williams, L.F.; and Green, D.E. 1964. An evaluation of damage to soybeans by brown and green stink bugs. *J. Econ. Entomol.* 57: 719-22.

ICRISAT. 1977. International Centre for Crops Research in the Semi-Arid Tropics Annual Report 1976/77. Hyderabad, India.

Kogan, J.; Sell, D.K.; Stinner, R.E.; Bradley, J.R., Jr.; and Kogan, M. 1978. The literature of arthropods associated with soybeans: 5, A bibliography of *Heliothis zea* (Boddie) and *H. virescens* (F.) (Lepidoptera: Noctuidae). INTSOY Series no. 17. 242 pp.

Lawn, R.J., and Byth, D.E. 1979. Soybean. In *Australian Field Crops,* ed. J.V. Lovett and Alec Lazenby, pp. 198-231. Sydney: Angus and Robertson.

Lawn, R.J., and Russell, J.S. 1978. Mungbeans: a grain legume for summer rainfall cropping areas of Australia. *J. Aust. Inst. Agric. Sci.* 44: 28-41.

McCarty, M.T.; Shepard, M.; and Turnipseed, S.G. 1980. Identification of predacious arthropods in soybeans using autoradiography. *Environ. Entomol.* 9: 199-203.

McWhorter, R., and Shepard, M. 1982. Predation by *Pristhesancus papuensis* on noctuid larvae. *Environ. Entomol.,* in press.

Miller, L.A.; Rose, H.A.; and McDonald, F.J.D. 1979. A sweep net survey of soybean insects, north west slopes and plains region, New South Wales, *Gen. and Appl. Entomol.* 11: 9-12.

Miner, F.D. 1966. Biology and control of stink bugs on soybeans. *Arkansas Agric. Exp. Sta. Bull.* no. 708. 40 pp.

Newsom, L.D. 1978. Progress in integrated pest management of soybean pests. In *Pest Control Strategies,* ed. E.H. Smith and D. Pimentel, pp. 157-80. New York Academic Press.

Passlow, T., and Waite, G.E. 1971. Green vegetable bug as a soybean pest. *Qld. Agric. J.* 97: 491-93.

Radunz, L. 1979. Working papers, Workshop on Tropical Agricultural Entomology, Mareeba, Queensland.

Richards, K.T. 1968. A study of the insect pest complex of the Ord River Irrigation Area. M.Sc. (Agric.) thesis, Univ. of West. Aust.

Sanders, G.J. 1978. Parasites of lepidopterous larvae in insect resistant and susceptible soybeans in South Carolina. M.S. thesis, Dept. of Entomology and Economic Zoology, Clemson Univ. 72 pp.

Shepard, M.; Carner, G.R.; and Turnipseed, S.G. 1974. A comparison of three sampling methods for arthropods of soybeans. *Environ. Entomol.* 3: 227-32.

______. 1975. Seasonal abundance of predaceous arthropods in soybeans. *Environ. Entomol.* 4: 985-88.

______. 1977. Colonization and resurgence of insect pests of soybeans in response to insecticides and field isolation. *Environ. Entomol.* 6: 501-6.

Spencer, K.A. 1977. A revision of the Australian Agromyzidae (Diptera). *West Aust. Museum Spec. Publ.* no. 8. 255 pp.

Teakle, R.E. 1973. A nuclear-polyhedrosis virus from *Heliothis punctigera* Wallengren (Lepidoptera: Noctuidae). *Qld. J. Agric. Anim. Sci.* 30: 161-77.

______. 1974. A granulosis virus from *Heliothis punctigera. J. Invert. Path.* 23: 127-29.

______. 1977. Disease organisms for pest control. *Qld. Agric. J.* 103: 1-4.

Turner, J.W. 1977. The nature of damage by *Nezara viridula* (L.) to soybean seed. *Qld. J. Agric. Anim. Sci.* 24: 105-7.

______. 1978. Pests of grain legumes and their control in Australia. In *Pests of Grain Legumes,* ed. S.R. Singh, H.F. van Emden, and T. Ajibola Taylor, pp. 73-81. London: Academic Press.

Turnipseed, S.G. 1972. Management of insect pests of soybean. *Proc. Tall Timbers Conf. on Ecology Animal Control and Habitat Management* 4: 189-203.

van Emden, H.F. 1981. Insects and mites of legume crops. In *Advances in Legume Science,* ed. R.J. Summerfield and A.H. Bunting, pp. 187-97. Royal Botanical Gardens, Kew.

Whitcomb, W.H. 1980. Sampling spiders in soybean fields. In *Sampling Methods in Soybean Entomology,* ed. M. Kogan and D.C. Herzog, pp. 544-58. New York: Springer-Verlag.

Williamson, A.J.P. 1971. Soybean is an expanding crop. *Qld. Agric. J.* 97: 571-79.

Wood, I.M. 1976. Preliminary results on the control of lucerne crown borer

(*Zygrita diva* Thomps.) in soybean crops in the Ord irrigation area. *Trop. Agron. Tech. Memo.* (CSIRO), no. 3. 11 pp.
Wood, I.M., and Russell, J.S. 1979. Grain legumes. In *Australian Field Crops,* ed. J.V. Lovett and Alex Lazenby, pp. 232-62. Sydney: Angus and Robertson.

INDEX 1: POTENTIAL PEST SPECIES

1. Reference: *Scientific and Common Names of Insects and Allied Forms Occurring in Australia* (Melbourne: CSIRO, 1980).
2. Where applicable, figure numbers are shown in bold type and text references are indicated by page numbers in ordinary type.

Family	Genus — Species	Common Name[1]	**Figure**/Page Reference[2]
COLEOPTERA			
Apionidae	*Apion* sp.	—	
Cerambycidae	*Corrhenes stigmatica* (Pascoe)	a lucerne crownborer	13
	Corrhenes sp.	a borer	13
	Paradisterna plumifera (Pascoe)	Speckled longicorn	13
	Zygrita diva Thomson	a lucerne crownborer	**1, 2, 3, 104, 112**/13
Chrysomelidae	*Aspidomorpha westwoodi* Boheman	a tortoise beetle	
	Aulacophora abdominalis (Fabricius)	Plain pumpkin beetle	**4**/12
	Aulacophora hilaris (Boisduval)	Pumpkin beetle	12
	Cryptocephalus parentheticus Suffrain	—	**5**/12
	Geloptera orientalis Lea	—	
	Monolepta australis (Jacoby)	Monolepta beetle, Redshouldered leaf beetle	**6**/12
	Nisotra submetallica Blackburn	—	
	Oulema sp.	—	
	Psylliodes sp.	a flea beetle	
	Rhyparida morosa Jacoby	Black leaf beetle	
Coccinellidae	*Henosepilachna vigintioctopunctata* (Fabricius)	Twentyeightspotted ladybird	**7**/12
Curculionidae	*Amnemus superciliaris* (Pascoe)	—	13
	Amnemus sp.	—	
	Graphognathus leucoloma (Boheman)	Whitefringed weevil	**8**/12
	Listroderes obliquus Klug	Vegetable weevil	
	Lixus mastersi Pascoe	Weed weevil	
	Onesorus conifer Lea	—	
	Onesorus 2 spp.	—	
	Perperus sp.	an apple root weevil	
	Synomus sp.	—	
Meloidae	*Zonitis* sp.	a blister beetle	
Tenebrionidae	*Gonocephalum* sp.	a false wireworm	14

DERMAPTERA				
	Labiduridae	*Nala lividipes* (Dufour)	Black field earwig	
DIPTERA				
	Agromyzidae	*Melanagromyza pseudograta* Spencer	—	**9**, **10**/11
		Melanagromyza sojae (Zehntner)	Soybean fly	13
		Ophiomyia phaseoli (Tryon)	Bean fly	**11**/13
HEMIPTERA				
	Aleyrodidae	*Trialeurodes vaporariorum* (Westwood)	Greenhouse whitefly	
	Alydidae	*Leptocorisa acuta* (Thunberg)	Paddy bug	9
		Melanacanthus scutellaris (Dallas)	Brown bean bug	**13**/9
		Mutusca brevicornis Dallas	—	
		Riptortus linearis (Fabricius)	—	9
		Riptortus serripes (Fabricius)	Podsucking bug	**14**/9
	Cercopidae	*Eoscarta carnifex* (Fabricius)	—	
	Cicadidae	Unidentified 2 spp.	cicadas	
	Cixiidae	*Olarius lubra* Kirkaldy	—	
	Colobathristidae	*Phaenacantha australiae* Kirkaldy	Linear bug	
	Coreidae	*Acroelytrum muricatum* Mayr	—	**15**/9
		Clavigralloides spinosus Dolling	—	
		Cletus sp.	—	
		Leptoglossus australis (Fabricius)	Passionvine bug	**16**/9
		Mictis profana (Fabricius)	Crusader bug	**17**/9
	Dictyopharidae	*Dictyophara* sp.	—	
	Eurybrachyidae	*Olonia* sp.	—	
	Flatidae	*Euphanta* sp.	—	**102**
	Lygaeidae	*Dieuches notatus* (Dallas)	—	
		Elasmolomus sordidus (Fabricius)	Peanut trash bug	
		Graptostethus servus (Fabricius)	—	**18**/9
		Horridipamera nietneri (Dohrn)	—	
		Melanerythrus mactans (Stål)	—	**19**/9
		Nysius clevelandensis Evans	Grey cluster bug	
		Nysius vinitor Bergroth	Rutherglen bug	9
		Spilosthethus decoratus (Stål)	—	
		Spilosthethus hospes (Fabricius)	—	
	Miridae	*Creontiades dilutus* (Stål)	Green mirid	**20**/9
		Taylorilygus pallidulus (Blanchard)	Brokenbacked bug	
		Trilaccus nigroruber Horvata	—	

Family	Genus — Species	Common Name	**Figure**/Page Reference
Pentatomidae	*Agonoscelis rutila* (Fabricius)	Horehound bug	**21**/9
	Dictyotus caenosus (Westwood)	Brown shield bug	**22**/9
	Eysarcoris trimaculatus Distant	—	**23**/9
	Glaucias amyoti (Dallas)	—	
	Jalloides sp.	—	
	Menestheus nervicus Dallas	—	
	Nezara viridula (Linnaeus)	Green vegetable bug	**24, 25, 99**/8, 9
	Oncocoris coelebs (Fabricius)	—	9
	Oncocoris hackeri McDonald	—	**26**/9
	Piezodorus hybneri (Gmelin)	Redbanded shield bug	**27, 109**/9
	Plautia affinis Dallas	Green stink bug	**28**/9
Pyrrhocoridae	*Dysdercus sidae* Montrouzier	Pale cotton stainer	
Rhopalidae	*Leptocoris mitellata* Bergroth	—	
	Liorhyssus hyalinus (Fabricius)	—	
Scutelleridae	*Coleotichus unicolor* Dallas	—	
	Philia senator (Fabricius)	—	
	Tectocoris diophthalmus (Thunberg)	Cotton harlequin bug	
Tessaratomidae	*Stilida indecora* (Stål)	—	
LEPIDOPTERA			
Arctiidae	*Amsacta marginata* (Donovan)	—	
	Utetheisa pulchelloides Hampson	Heliotrope moth	**29**/12
Gelechiidae	*Stomopteryx simplexella* (Walker)	Soybean moth	
Geometridae	*Gymnoscelis lophopus* Turner	—	**30**/11
	Hyposidra sp.	—	
Lycaenidae	*Catochrysops panormus platissa* (Herrich-Schaffer)	—	11
	Jamides phaseli (Mathew)	Bean flower caterpillar	**31**/11
	Lampides boeticus (Linnaeus)	Pea blue butterfly	11
	Zizina labradus (Godart)	Grass blue butterfly	**32**/11
Noctuidae	*Achaea janata* (Linnaeus)	Castoroil looper	**33, 34**/12
	Acontia nivipicta (Butler)	—	
	Acontia thapsina (Turner)	—	
	Agrotis sp.	a cutworm	
	Amyna octo Guenee	—	
	Anomis flava (Fabricius)	Cotton looper	12, 19
	Chrysodeixis argentifera (Guenee)	Tobacco looper	**35, 80, 84**/12, 18, 19
	Chrysodeixis eriosoma (Doubleday)	Green looper	**36, 37**/12, 18, 19
	Diachrysia orichalcea (Fabricius)	Soybean looper	12

		Eublemma dimidialis (Fabricius)	—	**38, 39**/10
		Eustrotia ritsemae (Snellen)	—	
		Heliothis armiger (Hubner)	Corn earworm, Cotton bollworm	**40**/10; **42, 74, 76, 79, 82, 83, 87-89, 115**/3, 15-19 (shared)
		Heliothis punctiger Wallengren	Native budworm	**41**/10; **42, 74, 76, 79, 82, 83, 87-89, 115**/3, 15-19 (shared)
		Mocis alterna (Walker)	—	**43, 44**/12
		Mythimna separata (Walker)	Northern armyworm	**45, 46**/12
		Saroba irrorata (Fabricius)	—	**47, 48**/12
		Spodoptera litura (Fabricius)	Cluster caterpillar	**49, 50, 81**/12, 16, 19
		Tarache thapsina (Turner)	—	
		Xanthodes congenita (Hampson)	Hairy leafeating caterpillar	
	Pterophoridae	*Sphenarches anisodactylis* (Walker)	—	
	Pyralidae	*Crocidolomia binotalis* Zeller	Cabbage cluster caterpillar	
		Etiella behrii (Zeller)	Lucerne seed web moth	**51**/10, 13
		Hedylepta diemenalis (Guenée)	—	
		Hymenia recurvalis (Fabricius)	Beet webworm	**52, 53**/12
		Loxostege massalis Walker	—	
		Maruca testulalis (Geyer)	Bean podborer	**54, 55, 85, 93, 94**/10, 16, 18
	Tortricidae	*Merophyas divulsana* (Walker)	Lucerne leafroller	
ORTHOPTERA				
	Acrididae	*Acrida conica* (Fabricius)	Longheaded grasshopper	
		Austracris guttulosa (Walker)	Spur-throated locust	**57**/11, 19
		Bermius odontocercus Stål	—	**58**/12
		Chortoicetes terminifera (Walker)	Australian plague locust	**59**/11
		Coryphistes 2 spp.	—	
		Gastrimargus musicus (Fabricius)	Yellow-winged locust	11
		Heteropternis obscurella (Blanchard)	—	
		Locusta migratoria (Linnaeus)	Migratory locust	11
		Oedaleus australis Saussure	—	
		Stenocatantops augustifrons (Walker)	—	
		Valanga irregularis (Walker)	Giant grasshopper	**60, 78**/11, 19
	Gryllidae	*Madasumma affinis* Chopard	—	
		Madasumma planiceps (Saussure)	—	**61**/12
		Metioche sp.	—	
	Pyrgomorphidae	*Attractomorpha similis* Bolivar	—	**62**/12
	Tettigoniidae	*Conocephalus laetum* Redtenbacher	—	
		Unidentified 1 sp.	a katydid	

INDEX 2: POTENTIAL BENEFICIAL SPECIES

1. Reference: *Scientific and Common Names of Insects and Allied Forms Occurring in Australia* (Melbourne: CSIRO, 1980).
2. Where applicable, figure numbers are shown in bold type and text references are indicated by page numbers in ordinary type.

	Family	Genus — Species	Common Name[1]	**Figure**/Page Reference[2]
		INSECTS		
COLEOPTERA				
	Cantharidae	*Chauliognathus nobilitatus* Erichson	a soldier beetle	15
		Chauliognathus pulchellus Macleay	a soldier beetle	**65**/15
	Carabidae	*Calosoma schayeri* Erichson	Green carab beetle	**66**/17
		Chlaenius flaviguttatus Macleay	—	**67**/17
		Geoscaptus sp.	—	
		Gnathaphanus pulcher Dejean	—	**68**/17
		Notogonum sp.	—	
		Pheropsophus verticalis Dejean	Bombardier beetle	
		Unidentified 1 sp.	—	
	Coccinellidae	*Coccinella repanda* Thunberg	Transverse ladybird	**70**/15
		Coelophora inaequalis (Fabricius)	Variable ladybird	
		Diomus notescens (Blackburn)	Minute twospotted ladybird	15
		Harmonia arcuata (Fabricius)	—	15
		Harmonia conformis (Boisduval)	Common spotted ladybird	**71**/15
		Micraspis aphidectoides Korschefsky	—	15
	Melyridae	*Dicranolaius bellulus* (Guérin-Méneville)	Pollen beetle	**73**/15
		Laius sp.	—	
DERMAPTERA				
	Labiduridae	*Labidura riparia truncata* Kirby	Common brown earwig	**74**/17
DIPTERA				
	Asilidae	*Bathypogon* sp.	a robber fly	
		Chrysopogon sp.	a robber fly	**75**/16
	Tachinidae	*Carcelia (?)cosmophilae* Curran	—	
		Carcelia sp.	—	
		Exorista sp.	—	
		Goniophthalmus australis (Baranov)	—	
		Palexorista sp.	—	**77**/19
		Peribaea (?)orbata (Wiedemann)	—	
		Peribaea sp.	—	

		Unidentified 1 sp. Tribe Blondelini	—	
		Unidentified 3 spp.	—	
HEMIPTERA				
	Lygaeidae	*Geocoris lubra* (Kirkaldy)	Bigeyed bug	16
	Miridae	*Campylomma livida* Reuter	Apple dimpling bug	16
		Tytthus chinensis (Stål)	—	
	Nabidae	*Tropiconabis nigrolineatus* (Distant)	a damsel bug	16
	Pentatomidae	*Cermatulus nasalis* (Westwood)	a predatory shield bug	**79**/16
		Oechalia schellenbergii (Guérin-Méneville)	a predatory shield bug	**80**/16
	Reduviidae	*Coranus triabeatus* Horrath	—	16
		Gminatus nigroscutellatus Breddin	—	**82**/16
		Penthikocleptes australicus Miller	—	
		Pristhesancus papuensis Stål	Assassin bug	**83**/16
		Sastrapada australica Stål	—	16
HYMENOPTERA				
	Braconidae	*Agathis rufithorax* Turner	—	
		Agathis sp.	—	
		Apanteles ruficrus (Haliday)	—	18
		Apanteles sp.	—	**84**/18
		Bracon 2 spp.	—	**85**/18
		Cremnops dissimilis Turner	—	**86**/18
		Cremnops 2 spp.	—	
		Cyanopterus sp.	—	
		Microplitis demolitor Wilkinson	—	**87-89**/18, 19
		Microplitis sp.	—	
	Chalcidae	*Brachymeria* 2 spp.	—	
		Dirhinus sp.	—	
	Encyrtidae	*Litomastix* sp.	—	**90**/18
	Eumenidae	*Anterhynchium nigrocinctum nigrocinctum* (Saussure)	—	**91**/16
		Epiodynerus tamarinus (Saussure)	—	**92**/16
		Pseudepipona bicolor (Saussure)	—	**93**/16
	Formicidae	*Iridomyrmex* sp.	a meat ant	**94**/16
	Halictidae	*Lasioglossum (Chilalictus)* sp.	—	
	Ichneumonidae	*Dicamptus* 2 spp.	—	
		Disophrys sp.	—	
		Heteropelma scaposum (Morley)	Twotoned caterpillar parasite	**95**/18, 19
		Lissopimpla excelsa (Costa)	Orchid dupe	**96**/18

Family	Genus — Species	Common Name	**Figure**/Page Reference
(Ichneumonidae) cont.	*Netelia producta* (Brullé)	Orange caterpillar parasite	**97**/18
	Temelucha cycnea Kerrich	—	
Mutillidae	*Ephutomorpha (?)morosa* Westwood	—	
	Ephutomorpha sp.	—	
Pompilidae	*Cryptocheilus* sp.	—	
	Ferreola handschini Haupt	—	**98**/16
Pteromalidae	*Eupteromalus* sp.	—	
	Perilampus sp.	—	
Scelionidae	*Trissolcus basalis* (Wollaston)	Green vegetable bug egg parasite	**99**/8
Sphecidae	*Ammophila* sp.	—	
	Chalybion sp.	—	
	Larra sp.	—	
	Liris 3 spp.	—	
	Podalonia suspiciosa Smith	—	
	Priononyx saevus (Smith)	—	**100**/16
Torymidae	*Syntomopus* sp.	—	
Vespidae	*Polistes townsvillensis austrinus* Richards	a papernest wasp	
	Polistes variabilis (Fabricius)	a papernest wasp	16
	Ropalidia gregaria (Saussure)	a papernest wasp	16
	Ropalidia revolutionalis (Saussure)	a papernest wasp	
	Ropalidia socialistica (Saussure)	a papernest wasp	
MANTODEA			
Mantidae	*Orthodera ministralis* (Fabricius)	Green mantid	**104**/17
	Tenodera sp.	—	
NEUROPTERA			
Chrysopidae	*Chrysopa* sp.	a green lacewing	**101**/17
ODONATA			
Aeshnidae	*Hemianax papuensis* (Burmeister)	—	
Coenagrionidae	*Austroagrion cyane* (Selys)	—	
	Ischnura aurora (Brauer)	—	**103**/16
	Ischnura heterosticta (Burmeister)	—	

Libellulidae	*Diplacodes bipunctata* (Brauer)	—	**102**/16
	Orthetrum caledonicum (Brauer)	—	
	Pantala flavescens (Fabricius)	—	

SPIDERS

ARANEIDA			
Araneidae	*Araneus heroine* (Koch)	an orbweaving spider	**105**/17
	Araneus theisi Walckenaer	an orbweaving spider	17
	Argiope protensa Koch	a teardrop spider	
	Argiope trifasciata (Forskoel)	an orbweaving spider	
Clubionidae	*Chiracanthium diversum* Koch	Yellow nightstalker	**106**/17
	Clubiona sp.	a sac spider	
Lycosidae	*Lycosa* 2 spp.	wolf spiders	**107**/17
Oxyopidae	*Oxyopes macilentus* Koch	a lynx spider	**108**/17
	Oxyopes molarius Koch	a lynx spider	**109**/17
	Oxyopes mundulus Koch	a lynx spider	
	Oxyopes 2 spp.	lynx spiders	
Pisauridae	*Dolomedes* sp.	Nursery web spider	
	Unidentified 1 sp.		
Salticidae	(?)*Astia* 2 spp.	jumping spiders	
	Cytaea sp.	a jumping spider	**110**/17
	Gangus longulus Simon	a jumping spider	
	Myrmarachne sp.	an ant-mimicking spider	
Sparassidae	*Olios* 2 spp.	badge huntsman spiders	
Theridiidae	*Achaearanea veruculata* (Urquhart)	a combfooted spider	
	Euryopis sp.	a combfooted spider	
	Latrodectus mactans hasselti Thorell	Redback spider	**111**/17
Thomisidae	*Diaea* sp.	a crab spider	**112**/17

DISEASES

	Entomophthora sp.		**113**/19
	Nomuraea rileyi		**114**/7, 19
	Nuclear polyhedrosis virus		**115**/19